Elastic Numbers

108 Puzzles for the Serious Problem Solver

Daniel Griller

Elastic Numbers

Proofread by Sam Hartburn (www.qbfproofreading.co.uk)

Cover design by Mariano Midaglia (www.2767web.com)

ISBN 978-0-9957826-0-0

For Mum, Dad and Anna

Preface

I composed the puzzles in this book over the course of three winters, and they first appeared in a maths competition I created at Hampton School in southwest London, where I have taught for the best part of five years.

I have a wonderful job; I cannot imagine another profession that would give me so much enjoyment day after day, and at Hampton I have been incredibly fortunate to work with some of the brightest young minds in the country. My teaching philosophy has always included showing pupils interesting mathematics they wouldn't get to see otherwise, such as Euclidean geometry, combinatorics and number theory. Hampton have been more than happy for me to share these mathematical curiosities with my pupils, and I am grateful to the school for accommodating my obsession.

Particular thanks go to Ben Bowles, Hampton's exceptional former Head of Mathematics, who took a chance by hiring me with no teaching experience, and who showed me unwavering support as my career began; my family, whom I have continuously bombarded with maths problems for all of my adult life; my problem testers, who spent a long time trialling the puzzles and provided me with critical feedback; my fellow department members, who make my working days happier than I could have asked for; and of course the pupils, without whom the competition (and these puzzles) would not have existed.

Daniel Griller, March 2017

How to use this book

If you like, you can dive straight in and see which problems take your fancy. Nevertheless, the following points should help you make the most of this book.

1. The puzzles have been arranged into four sections of increasing difficulty: Bronze, Silver, Gold and Diamond.

2. Do not be tempted to peek at the solutions too early! You will benefit far more from having given a problem serious thought than if you concede after only a few minutes, and you will become a better problem solver as a result.

3. The puzzles are designed to be solved with pen and paper only; no calculators!

4. It might help to use a ruler and compasses for some of the geometry problems, but be aware that diagrams are inaccurate, and you cannot solve a problem with measurement. You must use logical reasoning to arrive at the answer.

5. If, after reading a problem or solution, you would like something further explained, you can send in a query (details in Chapter 9).

Happy solving!

Contents

Part I

Problems

Chapter 1

Bronze

1. What is the value of $123 + 234 + 345$?

2. In a sale, a shopkeeper was supposed to take £5 off the original price of a toy. Instead, he divided the original price by 5. This resulted in a sale price of £3.50. What should the sale price have been?

3. What is the smallest three-digit multiple of 14?

4. What is $\frac{1}{2}$ of $\frac{1}{3}$ of $\frac{1}{4}$ of 168?

5. Every day for a week Penny baked some cakes. On the first day, she baked 5 cakes, and then on each day she baked twice as many cakes as she baked the day before. How many cakes did she bake in total?

6. What number is eight hundred and eighty eight less than one thousand one hundred and eleven?

7. Catherine has four tiles numbered 1 to 4. She pairs them up so that each pair is a two-digit prime. What is the difference between the two primes?

8. What is the value of $(2-4)-(6-8)$?

9. In the calculation $\boxed{3}\boxed{d}\boxed{9} \times \boxed{4} = \boxed{1}\boxed{5}\boxed{1}\boxed{6}$, what digit does d represent?

10. A film begins at 8.45pm. The film itself is split into four 25-minute sections, and the sections are separated by 5-minute advert breaks. At what time does the film end?

11. Olivia thinks of a number. She triples it, then adds 18, ending up with the number she first thought of. What was Olivia's number?

12. Alice buys 3 chocolate bars with a £5 note. She receives £3.23 in change. Then David buys 5 chocolate bars with a £5 note. How much change does he receive?

13. Emma thinks of a number. One fifth of her number is 6 more than one seventh of her number. What number is Emma thinking of?

14. The area of the shape below is 100. Find its perimeter.

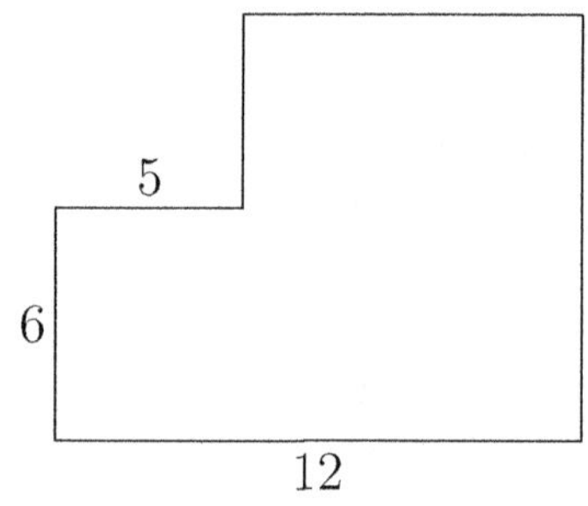

15. If $x = -1, y = -2$ and $z = -3$, what is the value of $\left(\left(x^2+y\right)^2+z\right)^2$?

16. James runs 50m north, then 150m east, then 250m south, then 450m west, then 200m north. How many metres is he from his starting point?

17. What fraction lies halfway between $\frac{1}{5}$ and $\frac{1}{6}$? *(Your answer should be in its simplest form.)*

18. In the sequence $3, 5, 8, 10, 13, \ldots$ we alternately add 2 then add 3 to get from one term to the next. The smallest multiple of 26 that appears is a and the smallest multiple of 27 that appears is b. Calculate $a+b$.

19. Grace divides 100 by each of the numbers 1, 2, 3, ..., 99, 100 and writes down the remainder in each case. For example, when dividing 100 by 2 she writes down 0, and when dividing 100 by 3 she writes down 1. What is the biggest remainder she writes down?

20. The figure below is made up of two rectangles and two triangles. What fraction of the total area is shaded? *(Give your answer in its simplest form.)*

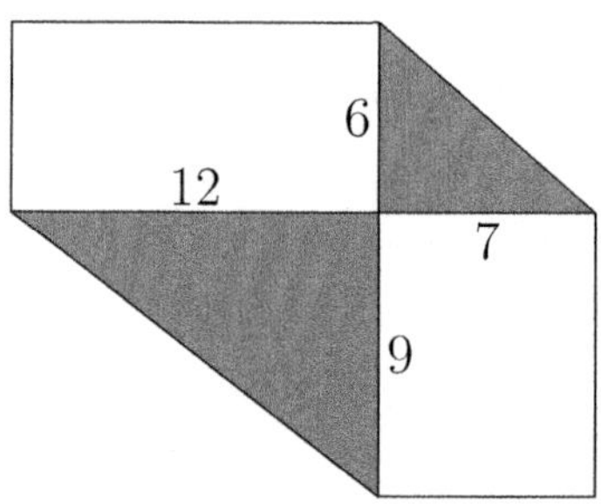

21. Alex has twice as many blue marbles as green marbles. He gives twenty of his blue marbles to Beatrice. Now he has three times as many green marbles as blue marbles.

    How many marbles does he now have?

22. How many cubes of side length 2 can be packed into a box with dimensions $14, 16$ and 18?

23. What is the smallest positive whole number that has two factors containing a digit 2, but does not contain a digit 2 itself?

24. Find the value of x in the regular pentagon below.

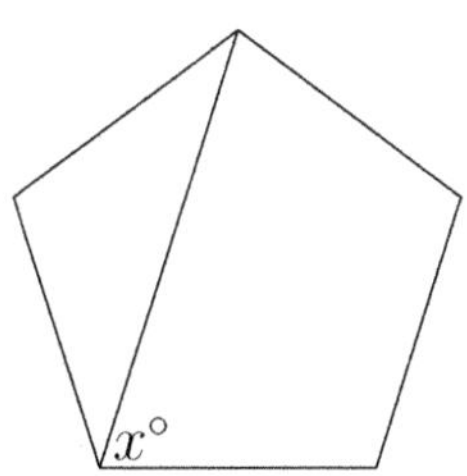

25. Suppose $xyz = 840$, where x, y and z are different positive whole numbers each containing the digit 2. Find the value of $x + y + z$.

26. Jess writes down a decimal number A, with two digits after the decimal point. She then rounds A to one decimal place to obtain a new number B. Finally she rounds A to the nearest whole number to obtain C. If $A + B + C = 20.17$, what is the value of A?

27. The symbol $\triangle$ is defined by $a \triangle b = 2a - 3b$. For example,

$4 \triangle 2 = 8 - 6 = 2$, and

$1 \triangle 3 = 2 - 9 = -7$.

Find the value of $(6 \triangle 7) \triangle 8$.

Chapter 2

Silver

1. Matthew arranges the two-digit factors of 1000 in ascending order. Which factor is in the middle of his list?

2. Calculate $\sqrt{33}$ to one decimal place.

3. Ares orders 30 pizzas for a party, each cut into eight slices. During the party, each of his guests eats the same number of slices. To his dismay, all the slices are gone before Ares has a chance to eat any.

   If each of the n guests ate m slices each, find the smallest possible value of $m + n$.

4. Jack and Sarah are taking part in a 40-mile cycling time trial. Jack completes the first 20 miles at an average speed of 20mph, and the last 20 miles at an average speed of 40mph. Sarah completes the first 20 miles at an average speed of 16mph. What average speed must Sarah achieve, in miles per hour, for the last 20 miles, to end with an identical finish time to Jack?

5. In the diagram, $AB = BD = DC$ and $\angle BAC = 76°$. Calculate the size, in degrees, of $\angle ABC$.

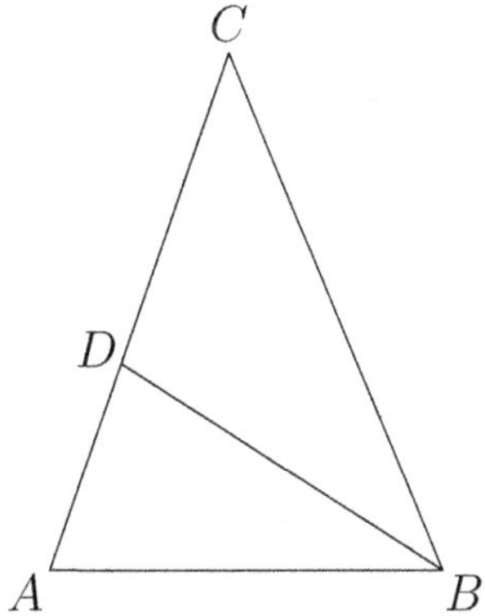

6. Sophie has just enough money to buy exactly 11 apples and 3 mangos, or exactly 2 apples and 5 mangos. Theo wants only one type of fruit, and has enough money to buy 1 mango, but not enough for 2, or enough money to buy x apples. Find the greatest possible value of x.

7. Write $\dfrac{4}{5 - \frac{6}{7}}$ as a single fraction in its simplest form.

8. Molly draws a brick pattern made up of 1×1 and 1×2 bricks as shown below. What is the total length of all the lines drawn?

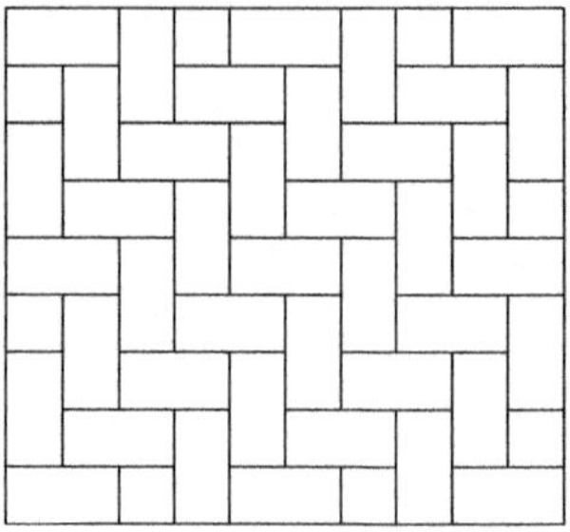

9. In Platoville, the only way to pay for anything is with 5p or 12p coins. Eric wishes to purchase an item for £1.65. If both he and the shopkeeper have a large number of 5p and 12p coins, what is the smallest number of coins that must change hands so that Eric pays £1.65 exactly?

10. If $(x-10)^2 = 9$ and $x^2 < 10^2$, what is the value of $(x+10)^2$?

11. A logo is designed by taking four squares of side length 5, drawing a diagonal line and two quarter circles, each of radius 5, and shading in various regions, as shown in the second diagram below. Calculate the total shaded area.

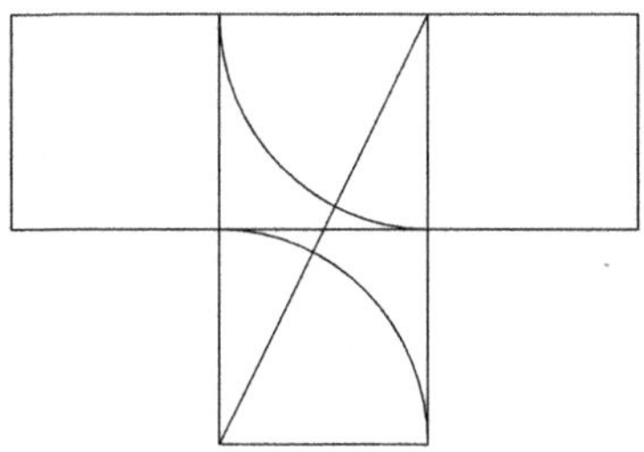

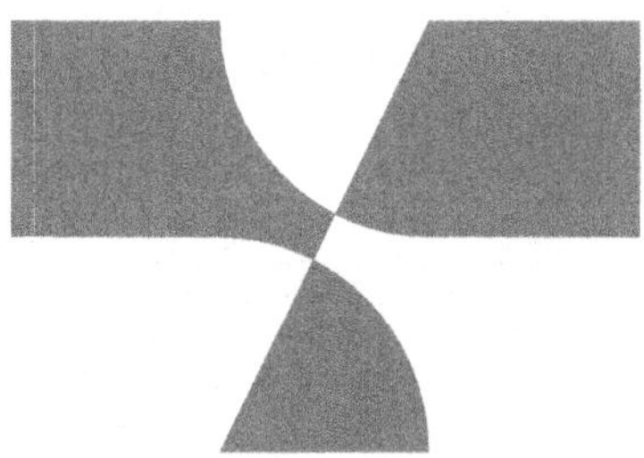

12. How many rearrangements of the letters in the word HOST have the T appearing somewhere to the left of the H?

13. Peter has 3 jumpers, 3 T-shirts and 3 pairs of trousers; one in each of the colours red, blue and yellow. How many different outfits containing a jumper, a T-shirt and a pair of trousers can he put together if he refuses to wear more than one blue item?

14. In the sequence 14, 18, 22, 26, ..., every term is four more than the previous term. How many of the first fifty terms are divisible by six?

15. The symbol $\diamond$ is defined by $a \diamond b = ab - \frac{a}{b}$. For example,

$6 \diamond 4 = 24 - 1.5 = 22.5$, and

$2 \diamond 2 = 4 - 1 = 3$.

If $x \diamond 5 = 24$, what is the value of x?

16. Hannah subtracts one prime number from another prime number and obtains a positive single digit number d. How many possible values can d take?

17. In the square $ABCD$ below, all five angles at the centre O are equal. Find the value of x.

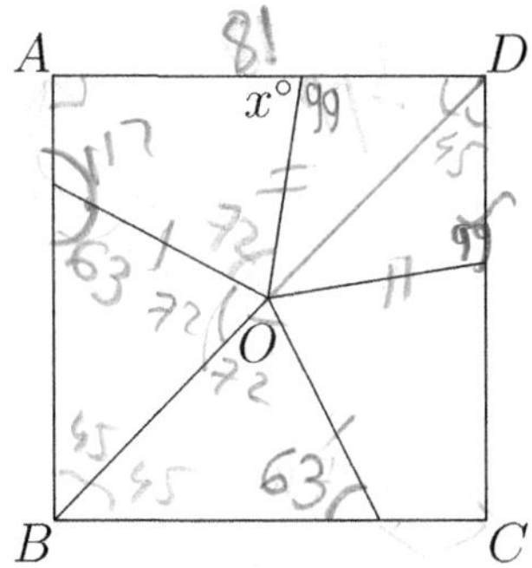

18. In the sum below, each of the letters A, B, C, D represents a different nonzero digit.

$$\begin{array}{ccc} & A & B \\ + & A & C \\ \hline B & D & A \end{array}$$

Determine, with proof, all possible solutions (A, B, C, D).

19. In rectangle $ABCD$ below, $BE = DE$ and angle $\angle EBC$ is three times angle $\angle ABD$. Calculate the size of angle $\angle DEB$.

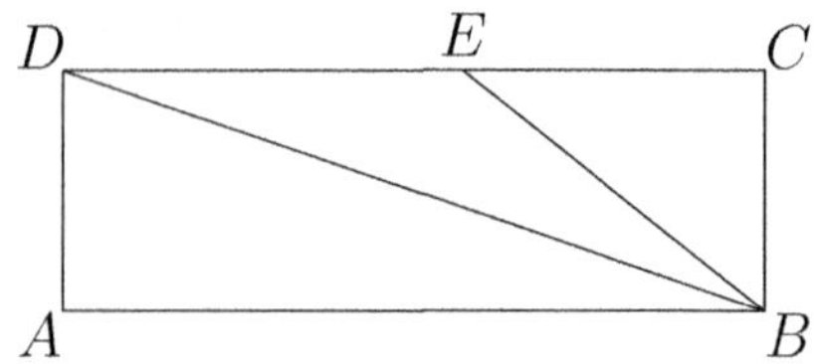

20. A *palindrome* is a sequence of letters that reads the same forwards and backwards. How many palindromes containing at least two letters can be formed using the letters in MANDARIN? *(The palindrome cannot contain any letter more often than it appears in the word MANDARIN.)*

21. I roll two standard dice, and *multiply* the two numbers together to obtain a total P.

    How many possible values of P are there?

22. 4 Jogs and 11 Rogs cost £15.00. 6 Jogs and 9 Rogs cost £18.30. How much do 7 Jogs and 8 Rogs cost?

23. N is a multiple of 6, 7 and 8. If N has three digits, what is the largest possible value of N?

24. The corners of an equilateral triangle with perimeter 30 are cut off, leaving a regular hexagon. What is the perimeter of the hexagon?

25. Three prime numbers p, q and r have a sum of 34. What is the *number* of possible values that p can take?

26. The grid below comprises five squares, each of side length 4. How many paths of length 36 begin at A and end at B, keeping to the grid lines at all times?

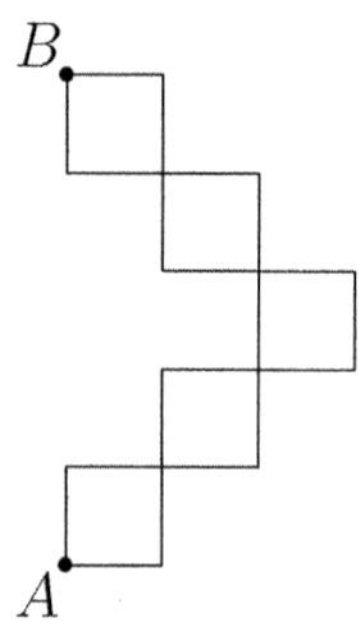

27. What is the smaller angle between the hands of a clock at 5:55?

Chapter 3

Gold

1. Six cards numbered 1 to 6 are placed in a line so that for each pair of neighbouring cards, one is a multiple of the other. In how many ways can this be done?

2. Nine equally spaced points on the circumference of a circle are joined up as shown below. Find the value of x.

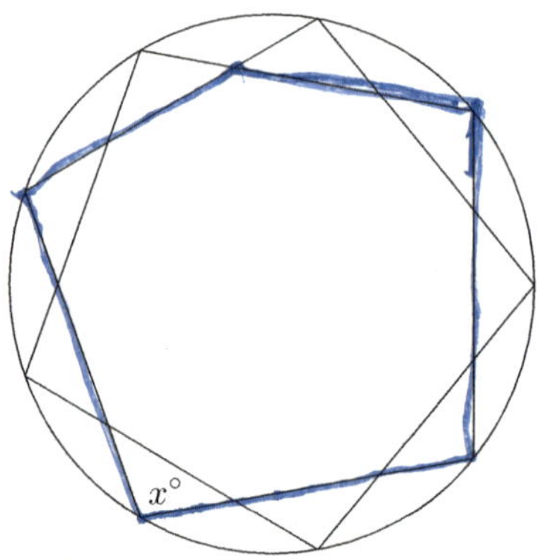

3. Tom is one year older than Lara. Tom is twice as old as Lara was when Lara was three times as old as Tom was twenty years ago. What is the sum of their ages now?

4. The square pattern below is symmetrical about both diagonals. $AC = 12$ and $DE = EF = 2$. Find the total area of the shaded region.

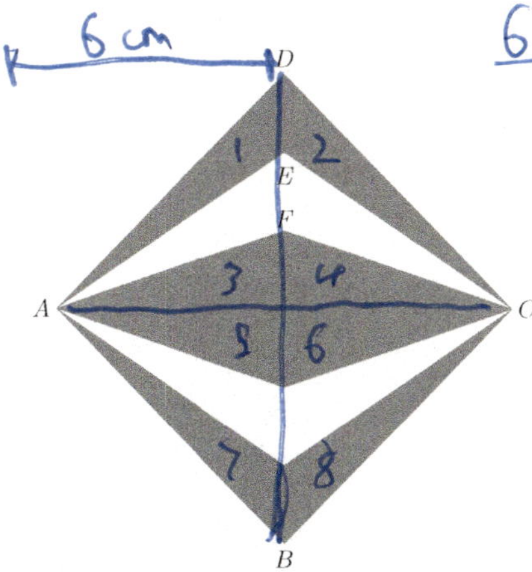

5. How many positive whole numbers less than 1000 can be written as the sum of four consecutive positive whole numbers?

6. Joseph creates a design. First he draws a large triangle. He then draws a second triangle by joining the midpoints of the sides of the first triangle. Then he draws a third triangle by joining the midpoints of the sides of the second triangle. He then draws a fourth triangle by joining the midpoints of the sides of the third triangle. Finally, he colours parts of his design as shown below.

Find the ratio of the area of the grey region to the area of the white region.

7. The first term of a sequence is 96, and thereafter each term is 2.5 times the previous term. What is the last whole number to appear in the sequence?

8. Define $\{x\}$ to mean 'x multiplied by $(x+1)$', for example

   $\{5\} = 5 \times 6 = 30$, and

   $\{-4\} = (-4) \times (-3) = 12$.

   If $\{k\} - \{k-1\} = 358$, what is the value of k?

9. Determine, with proof, the value of y in the regular decagon below.

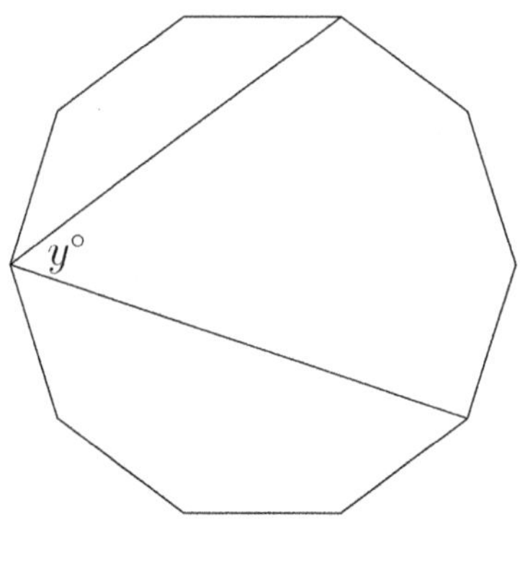

10. In the sequence of numbers $17, 3, 14, 11, 3, ...,$ every term after the first two terms is the difference of the two previous terms. Find the sum of the first 2015 terms of the sequence.

11. What is the smallest multiple of 36 in which the digits 1, 2 and 3 all appear?

12. In a triangle, one angle is double another and one angle is triple another. Find all possible numerical values (in degrees) of the smallest angle.

13. The perimeter and the area of the shape below (formed from five identical squares) are numerically equal. What is its area?

14. What is the largest whole number that is less than nine times its cube root?

15. Express $\frac{(1+2+3+\cdots+10)(3+6+9+\cdots+30)}{(2+4+6+\cdots+20)(4+8+12+\cdots+40)}$ as a decimal.

16. A rectangular sheet of paper is placed on top of another as shown. Given that the dimensions of the larger sheet are double those of the smaller sheet, find the total area covered by the sheets in this position.

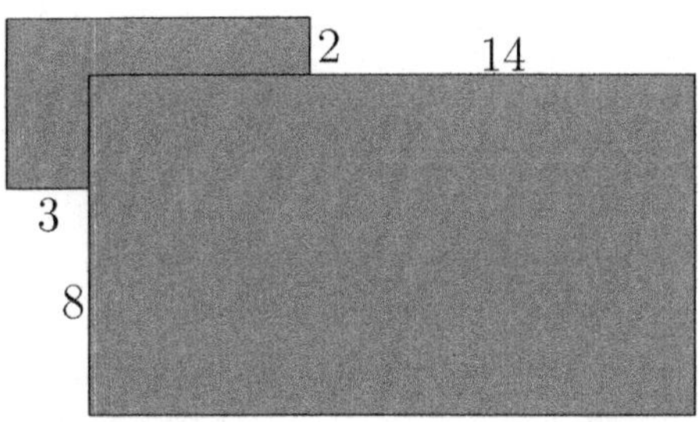

17. Joe scored 86% on a maths exam (rounded to the nearest %). What is the smallest possible number of marks he could have obtained?

18. Arthur, Bob and Charlie have 3 marshmallows between them. The following conversation takes place:

    Arthur: "I have more than half the marshmallows."

    Bob: "I have more marshmallows than Arthur."

    Charlie: "Exactly two of us are lying."

    Then Bob eats one of his marshmallows.

    How many marshmallows does Bob have left?

19. How many fractions of the form $\frac{1}{n}$ (where n is a positive whole number) are greater than 0.006 but less than 0.007?

20. Johnny draws a polygon with n edges so that each of its vertices is at one of the grid points below, and every interior angle is less than 180°. What is the largest possible value of n?

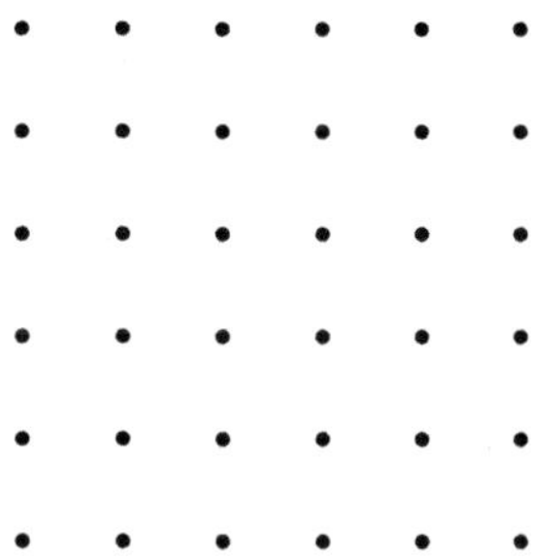

21. It takes 18 minutes to fill Bill's bathtub using both the hot and cold taps together. It takes 30 minutes using just the cold tap.

    How many minutes does it take using just the hot tap?

22. In the circle below, O is the centre, $\angle AOD = 90°$, $\angle OAB = 22°$ and AB is parallel to CD. Calculate $\angle ABC$.

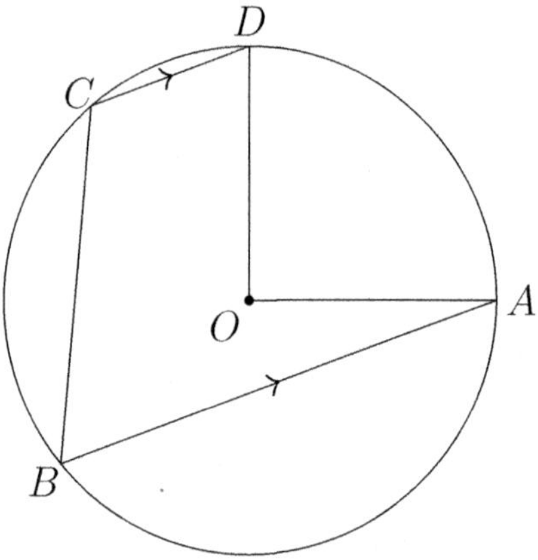

23. Raam and Tony live 3km apart, at opposite ends of a long road. At 2.00pm Raam leaves his house and starts walking at 4km/h along the road towards Tony's house. Ten minutes later, Tony leaves his house and starts walking at 6km/h towards Raam. At what time do they meet?

24. In a set of twenty one consecutive positive whole numbers, the sum of the first twelve numbers is equal to the sum of the last nine numbers. What is the largest number in the set?

25. Evariste is one of many performers standing in a rectangular grid formation, in which everyone is lined up in rows and columns. There are 175 people in all the rows in front of him combined, 400 in the rows behind him, 312 in the columns to his left, and 264 in the columns to his right.

If row 1 is at the front, and column 1 is on the far left, in which row and in which column is Evariste standing?

26. What is the smallest positive whole number n such that $75n$ is a cube?

27. Ashley starts listing numbers

$$1, 2, 3, 4, 6, 8, 9, 10, ...,$$

missing out every number containing a digit 5 or 7. What is the 1000th number in his list?

Chapter 4

Diamond

1. Phil chooses k different numbers from the set $\{1, 2, 3, 4, 5, 6, 7, 8, 9, 10\}$. No two of his choices add to a square number. Determine, with proof, the maximum possible value of k.

2. Consider the two sequences of dot figures below. In the first sequence, dots make the outline of a single square, with the side length of the square increasing by 1 each time.

In the second sequence, dots make the outline of two identical adjacent squares, with the side length of the squares again increasing by 1 every time.

Find the smallest whole number n greater than 2017, for which there is a figure in the first sequence and a figure in the second sequence, each containing exactly n dots.

3. We say a two-digit number "ab" is *elastic* if it has the following two properties:

   - Neither of the digits a or b is zero.
   - No matter how many zeros are put in between the digits a and b, the number "$a000\cdots00b$" is always divisible by the two-digit number "ab".

   Find three elastic numbers, and explain why each of them is elastic.

4. Euclid United and Sphere City played a football match. Sphere City won 9-8, and at no point was the goal difference more than a single goal. Given this information, one possible goal sequence was ESSESESEESESESSES (Euclid scored once, then Sphere scored twice, then Euclid scored once, and so on...).

   How many possible goal sequences were there for this match?

5. Anna and Daniel play a game. Starting with 30 beans on a table, the two players take turns removing a number of beans. Anna may remove 2 or 5 beans on her turn, and Daniel may remove 3 or 4 beans on his. The first player who cannot move loses the game, and the other player is the winner.

   Which player can guarantee a win if Anna goes first?

6. Suppose n is a positive whole number such that $3n$ has exactly 5 positive factors (including 1 and itself). How many positive factors does $4n$ have?

7. Helen is on an island. There are six trees located at various points on the island, and all the trees are identical. Every day for a week, she sets up camp at a different point on the island, looks all around her and records the number of trees she can see (a tree will be hidden from view precisely when it is directly behind another tree). After a week she has written down T different numbers. Find the greatest possible value of T, and prove that your answer is correct.

8. How many positive whole numbers less than 1000 leave the same remainder when divided by 3, 4 or 5?

9. A square has vertices at the points $(1, 6)$, $(-10, -5), (1, -16)$ and $(12, -5)$. The x and y axes split the square into two quadrilaterals, a triangle and a pentagon. Find the area of the pentagon.

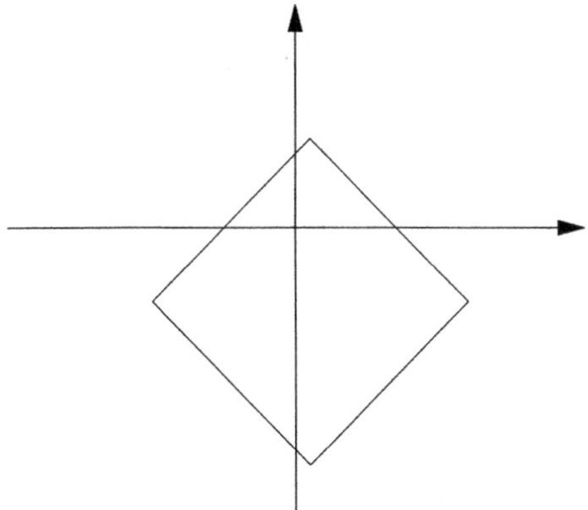

10. Arun splits the numbers $\{2, 3, 4, 5, 6\}$ into two groups, with at least one number in each group. He then adds the lowest common multiple of the first group to the lowest common multiple of the second group, to obtain a total T. What is the difference between the largest and smallest possible values of T?

11. Call a whole number *cool* if it is one more than a multiple of 3. What is the sum of all the cool numbers between -300 and 300?

12. Prove that every odd number (positive or negative) lies exactly halfway between a multiple of 3 and a multiple of 4.

13. Every pupil at the Euler Academy studies French or Spanish. At the start of the year, one third of the French students also studied Spanish, and two fifths of the Spanish students also studied French. After one term, six of the double-linguists dropped French, so that now only a quarter of the French students study Spanish. How many pupils are at the Euler Academy?

14. In the diagram below, $ABCD$ and $DEFG$ are squares. Prove that $\triangle ADG$ and $\triangle CDE$ have the same area.

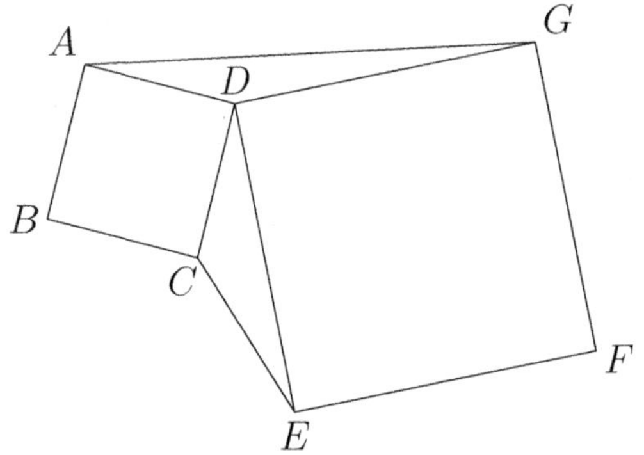

15. A toy company has two factories. Factory A has 20 workers, and each worker can make a robot in 2 minutes, or a stuffed animal in 3 minutes. Factory B has 24 workers, and each worker can make a robot in 3 minutes, or a stuffed animal in 4 minutes. In one hour the company needs to produce an equal number of robots and stuffed animals. If each worker must work for the entire hour, and is only allowed to make one type of toy, what is the most toys the company can produce?

16. How many five-digit numbers contain an even digit and an odd digit?

17. Mr and Mrs Newton and their two children meet Mr and Mrs Leibniz and their two children at the cinema. In how many ways can the eight of them arrange themselves in a single row of eight seats, so that each of them is sat next to a member of their own family?

18. How many triangles have all of their vertices at points on the cross below?

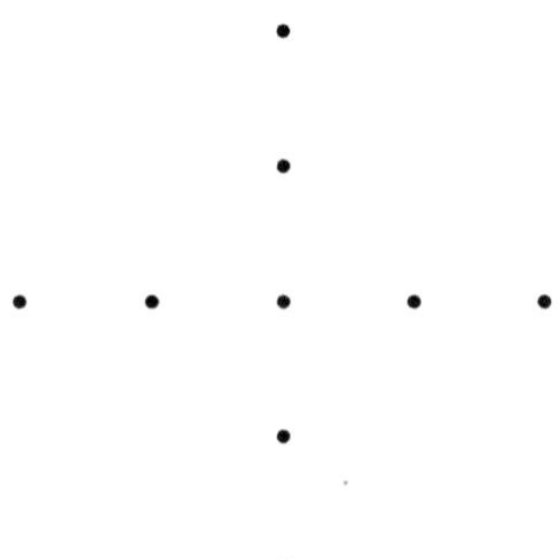

19. How many positive multiples of $10^3 - 1$ are factors of $10^6 - 1$? *(Every whole number is a multiple/factor of itself.)*

20. Ashley thinks of a four-digit multiple of 4. When its digits are reversed, he obtains another four-digit multiple of 4. Find the difference between the maximum and minimum possible values of Ashley's original number.

21. In the semicircle below, O is the centre and AB is parallel to OC.

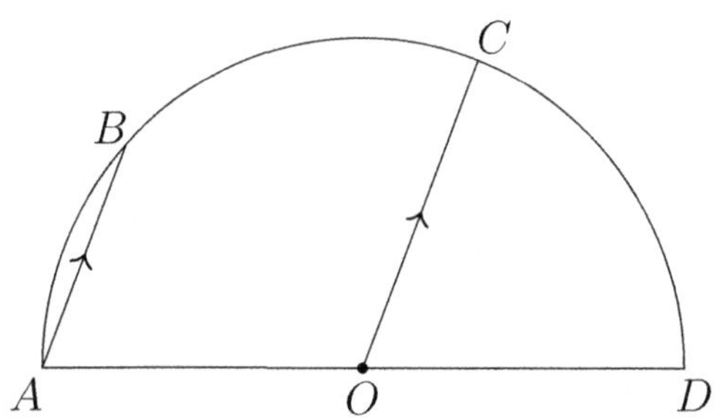

Prove that the distances BC and CD are equal.

22. We say a positive whole number is *solar* if its smallest prime factor is 5. How many positive whole numbers less than 2016 are solar?

23. A screen displays a number. Pressing button X subtracts 1 from the number displayed, and pressing button Y multiplies the number displayed by -2. Starting with the number n, Clare presses button X, then Y, then X, then Y, and so on, until each button has been pressed ten times. The final number is also n.

    What is the value of n?

24. When Anthony and Benjamin run round a circular track in the same direction at constant but different speeds, they meet every 3 minutes. When Benjamin changes direction (but maintains his speed) they meet every 40 seconds.

    If Anthony is faster than Benjamin, calculate

$$\frac{\text{speed of Anthony}}{\text{speed of Benjamin}}.$$

25. In how many ways can a class of six girls and six boys be split into two teams, with three girls and three boys on each team?

26. In the triangle below, CN and MB are straight lines, $\angle CAB = 90^\circ$ and $CM = MA = AN = NB = 5$. Find the exact area of the shaded region.

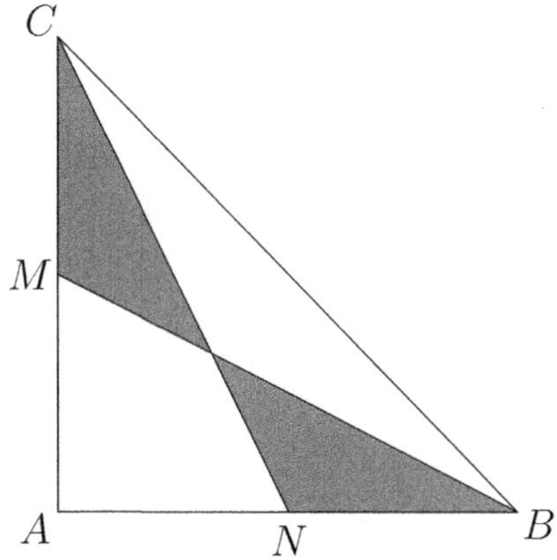

27. I have a large supply of red square tiles and blue square tiles. How many different 3×3 square patterns can I make using nine of these tiles, if two patterns are considered identical whenever one is a rotation of the other?

Part II

Solutions

Chapter 5

Bronze Solutions

1. $123 + 234 + 345 = \boxed{702}$

2. The original price was £3.50 × 5 = £17.50, so the sale price should have been £17.50 − £5 = $\boxed{£12.50}$

3. $14 \times 7 = 98$ and $14 \times 8 = 112$, so the answer is $\boxed{112}$

4. $\frac{1}{2} \times \frac{1}{3} \times \frac{1}{4} \times 168 = \boxed{7}$

5. In total, Penny bakes $5 + 10 + 20 + 40 + 80 + 160 + 320 = \boxed{635}$ cakes.

6. $1111 - 888 = \boxed{223}$

7. The two primes are 41 and 23, so the difference is $41 - 23 = \boxed{18}$

8. $(2-4)-(6-8)=(-2)-(-2)=\boxed{0}$

9. $1516 \div 4 = 379$, so $d = \boxed{7}$

10. From start to finish, the film (including advert breaks) lasts $25+5+25+5+25+5+25$ minutes, that is 1 hour and 55 minutes. So the film ends at $\boxed{10.40\text{pm}}$

11. Let x be Olivia's number. We have $3x+18=x$, so $2x=-18$ and $x=\boxed{-9}$

12. Alice spent £1.77, so each chocolate bar costs 59p. Hence David spent £2.95, and he received $\boxed{£2.05}$ in change.

13. Let Emma's number be x. Then we have $\frac{x}{5}=6+\frac{x}{7}$, so $7x=210+5x$, so $2x=210$ and $x=\boxed{105}$

14. Split the diagram into two rectangles as shown.

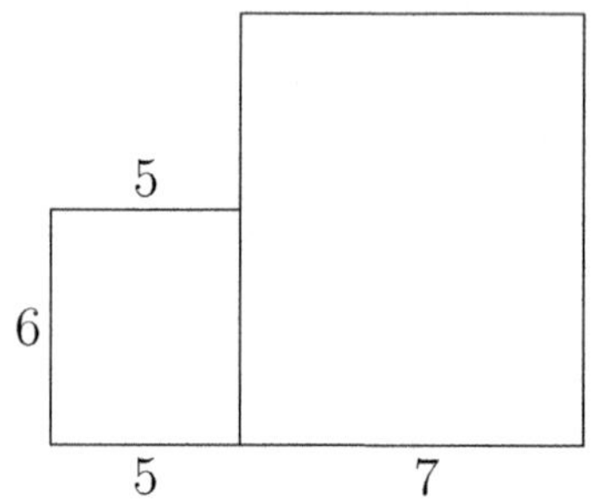

The area of the left rectangle is 30, so the area of the right rectangle must be $100 - 30 = 70$. Since its width is 7, its height must be 10. Thus the perimeter of the entire shape is

$12 + 10 + 7 + 4 + 5 + 6 = \boxed{44}$

15. $x^2 = (-1)^2 = 1$, so
$x^2 + y = 1 + (-2) = -1$, so
$(x^2 + y)^2 = 1$, so
$(x^2 + y)^2 + z = 1 - 3 = -2$, so
$((x^2 + y)^2 + z)^2 = (-2)^2 = \boxed{4}$

16. In total, James runs 250m north, 250m south, 150m east and 450m west. So he ends up exactly 300m west of his starting point, and the answer is $\boxed{300}$

17. $\frac{1}{5} = \frac{12}{60}$ and $\frac{1}{6} = \frac{10}{60}$, so the fraction halfway between them is $\boxed{\frac{11}{60}}$

18. The sequence contains every positive whole number whose last digit is $0, 3, 5$ or 8. The smallest multiple of 26 with this property is 78, and the smallest multiple of 27 with this property is 108. Hence the answer is $78 + 108 = \boxed{186}$

19. When 100 is divided by 51, the remainder is 49. The remainder is always smaller than the number we divide by, so to obtain a remainder of 50 or greater, we would need to divide 100 by 51 or greater. But each of these divisions produce a remainder of less than 50. It follows that the largest possible remainder is $\boxed{49}$

20. The small triangle has area $\frac{1}{2} \times 6 \times 7 = 21$, the big triangle has area $\frac{1}{2} \times 12 \times 9 = 54$, the higher rectangle has area $6 \times 12 = 72$ and the lower rectangle has area $7 \times 9 = 63$, so the required fraction is $\frac{21+54}{21+54+72+63} = \frac{75}{210} = \boxed{\frac{5}{14}}$

21. Let B and G be respectively the number of blue marbles and the number of green marbles Alex has at the beginning. The information given leads to $B = 2G$ and $G = 3(B - 20) = 3B - 60$, so $G = 6G - 60$, so $5G = 60$. Hence $G = 12$ and $B = 24$. Thus he now has $12 + 4 = \boxed{16}$ marbles.

22. You can fill the box completely with 7, 8 and 9 boxes along the different edges. In total this is $7 \times 8 \times 9 = \boxed{504}$ cubes.

23. The first few numbers containing the digit 2 are $2, 12, 20, 21, \ldots$. Note that 36 is a multiple of both 2 and 12, but doesn't contain a digit 2 itself. Since $2 \times 20 = 40$ is already bigger than 36, any other candidate would have to be divisible by 12. But the only other possibility is 24, which contains a digit 2. Thus the answer is $\boxed{36}$

24. Consider the diagram below.

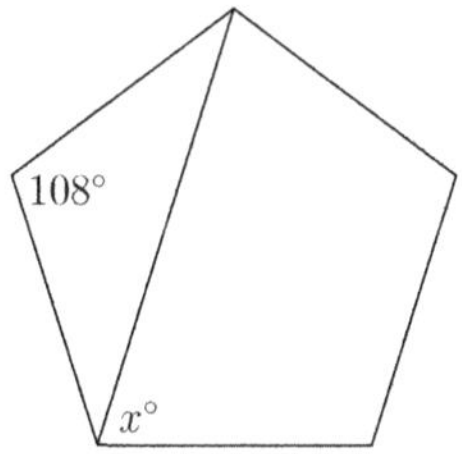

The sum of the interior angles of a pentagon is $(5-2) \times 180° = 540°$, so each interior angle in a regular pentagon is 108°. The triangle in the diagram is isosceles, so each of its base angles must equal 36°, and so the value of x is $108-36 = \boxed{72}$

25. $840 = 2 \times 420 = 2 \times 20 \times 21$, and there is no other way to write 840 as the product of three numbers each containing the digit 2. So $x + y + z = 2 + 20 + 21 = \boxed{43}$

26. We see that $6.57 + 6.6 + 7 = 20.17$, and increasing/decreasing the leftmost number will just increase/decrease the number on the right hand side of the equation. Thus $A = \boxed{6.57}$

27. $(6 \triangle 7) \triangle 8$

$$= (12 - 21) \triangle 8$$
$$= (-9) \triangle 8$$
$$= -18 - 24$$
$$= \boxed{-42}$$

Chapter 6

Silver Solutions

1. The two-digit factors of 1000 are $10, 20, 25, 40$ and 50, so the factor in the middle of Matthew's list is $\boxed{25}$

2. $5.7^2 = 32.49$, $5.8^2 = 33.64$ and $5.75^2 = 33.0625$, so the answer is $\boxed{5.7}$

3. There were $30 \times 8 = 240$ slices in total, so $mn = 240$. The smallest value of $m + n$ occurs when $m = 16$ and $n = 15$ (or the other way round), when m and n are close together, and so the answer is $\boxed{31}$

4. Jack takes 1 hour for the first 20 miles, then half an hour for the second 20 miles; that is 1.5 hours in total. Sarah takes 1.25 hours for the first 20 miles, so must take only 0.25 hours for the second 20 miles. This requires an average speed of $\boxed{\text{80mph}}$

5. Consider the diagram below.

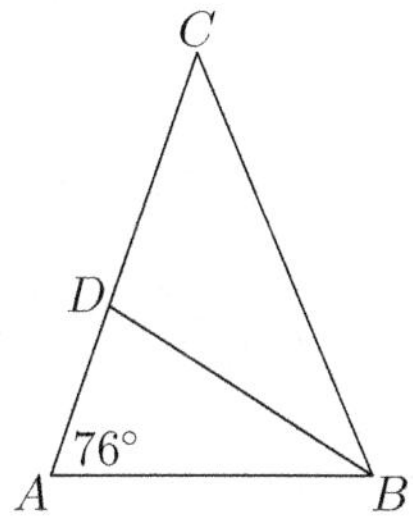

$\triangle ABD$ is isosceles, so $\angle ADB = \angle BAD = 76°$, and $\angle ABD = 28°$. So $\angle BDC = 104°$. But $\triangle BDC$ is also isosceles, so $\angle DBC = \angle DCB = 38°$. Thus $\angle ABC = \boxed{66°}$

6. Sophie can trade 9 apples for 2 mangos. Theo has enough for 1 mango, but not 2, so he doesn't have enough for 9 apples, but he might have enough for 8. So the maximum possible value of x is $\boxed{8}$

7. $\frac{4}{5-\frac{6}{7}} = \frac{4}{\frac{35}{7}-\frac{6}{7}} = \frac{4}{(\frac{29}{7})} = \frac{4\times 7}{29} = \boxed{\frac{28}{29}}$

8. The dimensions of the pattern are 9-by-10. Consider a 9-by-10 grid split up into 1-by-1 squares.

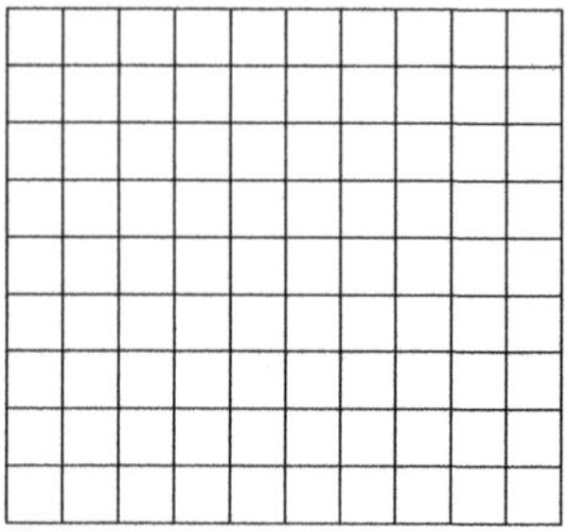

In this case the total length of all lines drawn would be $10 \times 10 + 9 \times 11 = 199$. To answer the problem, we simply have to subtract 1 for every 2-by-1 brick that appears in the original pattern. There are 21 horizontal bricks and 20 vertical bricks, so the total length of all lines drawn is $199 - 21 - 20 = \boxed{158}$

9. Eric can pay £1.65 exactly using ten 12p coins and nine 5p coins, and this uses 19 coins in total. Reducing the number of 12p coins will increase the number of 5p coins, and since 5p is less than 12p, this will only cause more coins to change hands. He could pay £1.80 with fifteen 12p coins, and receive three 5p coins in change, for 18 coins in total. Now any increase in the number of 12p coins will just increase the number the coins received in change. Hence the answer is $\boxed{18}$

10. $(x - 10)^2 = 9$, so $x - 10 = 3$ or -3. So $x = 13$ or 7. But $x^2 < 10^2 = 100$, so $x = 7$, and $(x + 10)^2 = 17^2 = \boxed{289}$

11. Consider the diagram below.

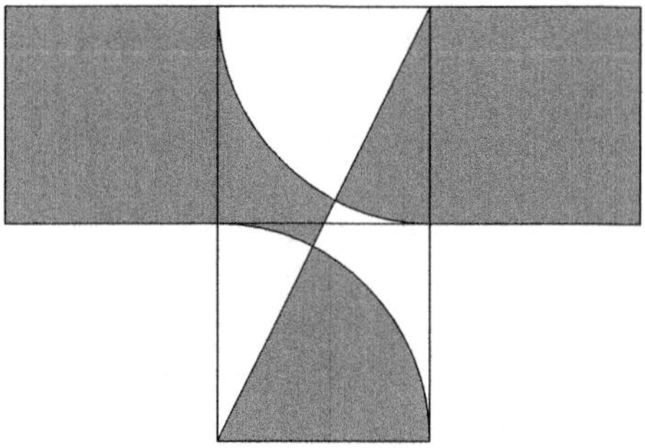

The regions can be rearranged to form three 5-by-5 squares.

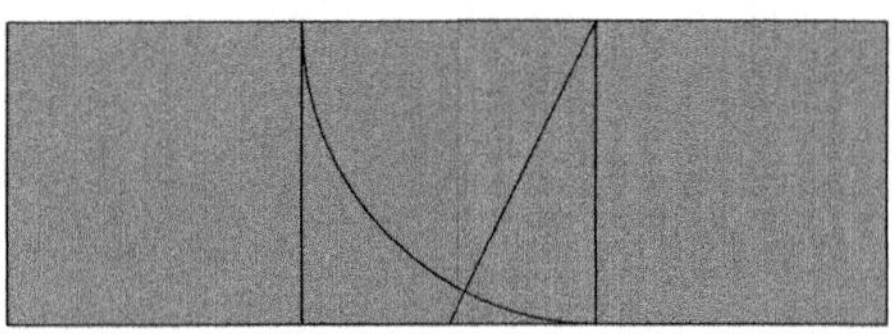

Thus the total shaded area is simply
$3 \times 5 \times 5 = \boxed{75}$

12. Of the $4 \times 3 \times 2 \times 1 = 24$ arrangements of the letters in the word HOST, exactly half of them have the T appearing somewhere to the left of the H, so the answer is $\boxed{12}$

13. With no blue items, Peter has $2 \times 2 \times 2 = 8$ choices of outfit. With one blue item, he has 3 choices for the blue item of clothing, then he has $2 \times 2 = 4$ choices for the colours of the other two items. That is 12 possible outfits with one blue item. So the number of possible outfits is $8 + 12 = \boxed{20}$

14. The multiples of six occur at positions 2, 5, 8, ..., 50; that is $\boxed{17}$ multiples of six in total.

15. $x \diamond 5 = 24$, so $5x - \frac{x}{5} = 24$, so $25x - x = 120$, so $24x = 120$. Thus $x = \boxed{5}$

16. The first few primes are $2, 3, 5, 7, 11$. We have $3-2 = 1$, $5-3 = 2$, $5-2 = 3$, $7-3 = 4$, $7-2 = 5$, $11-5 = 6$, $11-3 = 8$ and $11-2 = 9$. So d could equal any of $1, 2, 3, 4, 5, 6, 8, 9$. For d to equal 7, we would require one even prime and one odd prime. The only even prime is 2, however 9 is not prime. Hence $d = 7$ is impossible, and the answer is $\boxed{8}$

17. Consider the diagram below.

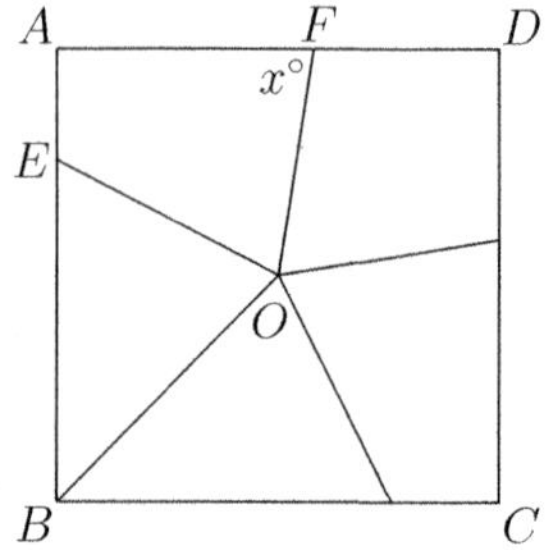

$\angle BOE = 72°$ and $\angle EBO = 45°$, so
$\angle BEO = 63°$ and $\angle AEO = 117°$.
Also, $\angle EOF = 72°$ and $\angle EAF = 90°$. Finally, using the fact that angles in a quadrilateral sum to $360°$, we obtain $x = 360 - 90 - 117 - 72 = \boxed{81}$

18. Since the sum of 2 two-digit numbers is at most 198, $B = 1$. We also require $A = 5, 6, 7, 8$ or 9 to force a carry into the third column. If $C = 9$ then from the rightmost column we obtain $A = 0$, which is no good. So $C < 9$ and there is no carry into the middle column. Considering each of the possibilities for A in turn we obtain the sums $51 + 54 = 105, 61 + 65 = 126, 71 + 76 = 147, 81 + 87 = 168$ and $91 + 98 = 189$. The first of these fails since $D = 0$ and the last fails as $C = D$. Hence the complete set of solutions is

$$\boxed{(A, B, C, D) = (6, 1, 5, 2), (7, 1, 6, 4), (8, 1, 7, 6)}$$

19. Consider the diagram below.

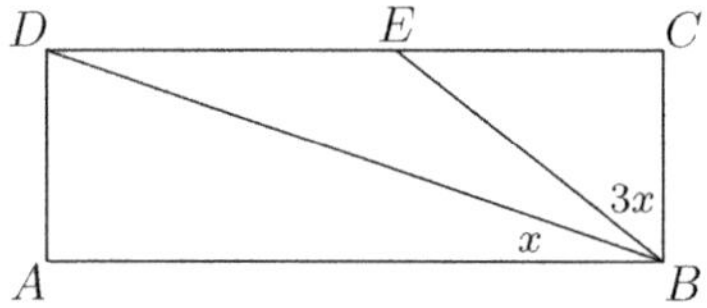

Let $\angle ABD = x$, so that $\angle EBC = 3x$. By alternate angles, $\angle BDE = x$. Since $\triangle BDE$ is isosceles, $\angle DBE = x$. Given that $\angle ABC$ is a right angle, we now have $5x = 90°$, so $x = 18°$. Hence the base angles in $\triangle BDE$ are both $18°$, and thus the required angle is

$180 - 18 - 18 = \boxed{144°}$

20. In alphabetical order, the letters we have available are AADIMNNR. There are 2 two-letter palindromes (AA, NN). For three-letter palindromes, we have 2 options for the first letter, and 5 choices for the middle letter, so 10 choices in total. For four letters, we have only two options (ANNA, NAAN). For five letters, we have two choices for the four outside letters (AN...NA, NA...AN) and 4 choices for the middle letter, so 8 choices in total. Since only A and N are repeated, there are no palindromes containing six letters or more. Hence the answer is
$2 + 10 + 2 + 8 = \boxed{22}$

21. The possible values of P are $1, 2, 3, 4, 5, 6, 8, 9, 10, 12, 15, 16, 18, 20, 24, 25, 30$ and 36, that is $\boxed{18}$ possible values.

22. Considering the change in price from the first scenario to the second, we see that 2 Jogs cost £3.30 more than 2 Rogs. So 1 Jog costs £1.65 more than 1 Rog, and hence the required price is $£18.30 + £1.65 = \boxed{£19.95}$

23. Since N is divisible by $6, 7$ and 8, it must be divisible by their lowest common multiple 168. The largest multiple of 168 that has only three digits is $\boxed{840}$

24. Consider the diagram below.

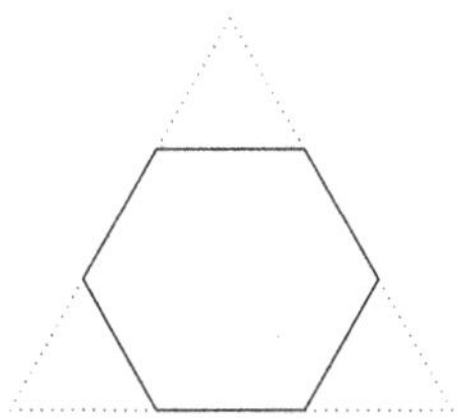

The corners that are removed are all equilateral triangles, whose side length is the same as that of the hexagon. Each side of the triangle has length 10, so each side of the hexagon has length $\frac{10}{3}$. Hence the perimeter of the hexagon is $6 \times \frac{10}{3} = \boxed{20}$

25. Since 34 is even, and the only even prime is 2, exactly one of p, q and r is even, and the other two are odd. So one of them is 2, and the others add to 32, so could be 3 and 29 or 13 and 19. Hence $p = 2, 3, 13, 19$ or 29; that is $\boxed{5}$ possible values.

26. Nine sides must be traversed in total, which means that we must take the short route through the middle square. For each of the other four squares, we have two possible paths, so the total number of paths is $2 \times 2 \times 2 \times 2 = \boxed{16}$

27. It is useful to note that there are 30° between neighbouring clock face numbers. At 6:00, the angle between the hands is 180°.

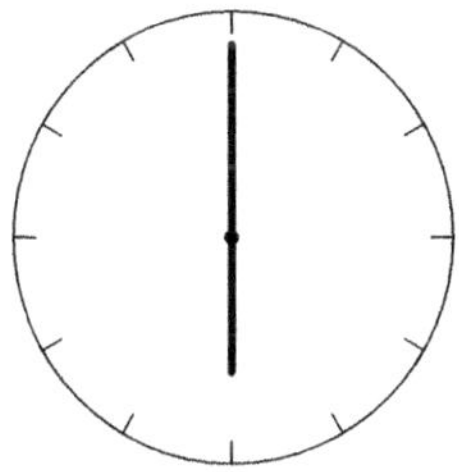

The hour hand moves through 0.5° per minute, so at 5:55, the hour hand is pointing between the numbers 5 and 6, and is 2.5° from the 6. At the same time, the minute hand is pointing at the number 11.

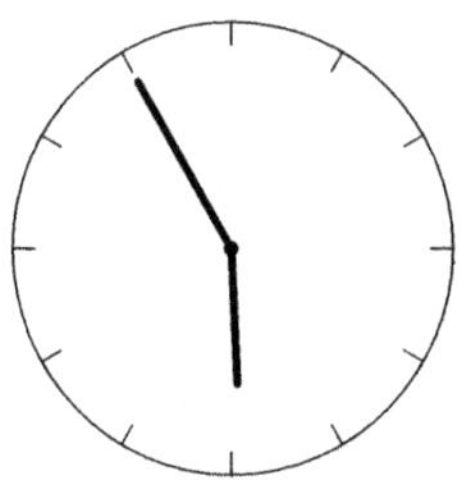

It follows that at 5:55, the angle between the hands is 152.5°

Chapter 7

Gold Solutions

1. The 5 must be placed at one end, next to the 1, and the 3 can only be placed next to the 1 or the 6. This leaves $(5, 1, 3, 6, 2, 4)$, $(4, 2, 6, 3, 1, 5)$, $(5, 1, 4, 2, 6, 3)$ and $(3, 6, 2, 4, 1, 5)$ as the only possibilities, so the answer is $\boxed{4}$

2. Imagine walking round the star, beginning at one of the vertices on the circumference, and following the straight edge as far as you can, until you reach another point on the circumference of the circle. Then you turn and follow the next edge, and repeat until you return to the starting point. By the end, you have turned through two complete revolutions, or 720°, and you did this by making nine equally sized turns at each of the nine vertices on the circumference. So at each vertex you turned through $\frac{720}{9} = 80°$.

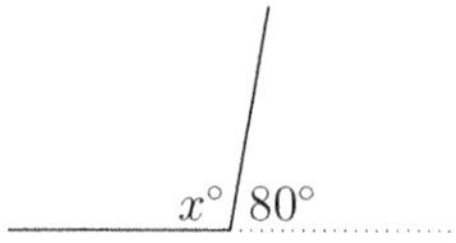

Hence $x = 180 - 80 = \boxed{100}$

3. Let T and L be respectively the present ages of Tom and Lara. Then $T = L + 1$. Twenty years ago Tom was aged $T - 20$, and when Lara was three times this age, she was $3(T - 20)$ years old. It follows that $T = 2 \times 3(T - 20) = 6(T - 20) = 6T - 120$, so that $5T = 120$ and $T = 24$. Thus $L = 23$ and the required sum is $24 + 23 = \boxed{47}$

4. Drawing in both diagonals AC and BD splits the figure into 12 triangles with the same base and same height (and so they all have equal areas).

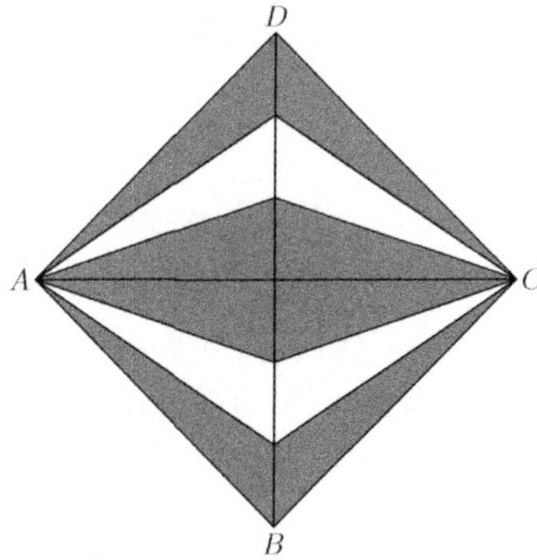

Eight of the triangles are shaded, so two thirds of the figure is shaded. Since the total area is $\frac{1}{2} \times 12 \times 12 = 72$, the area of the shaded region is $\boxed{48}$

5. The smallest such number is $1 + 2 + 3 + 4 = 10$ and the largest is $248 + 249 + 250 + 251 = 998$, so there are $\boxed{248}$ such numbers.

6. Consider the diagram below.

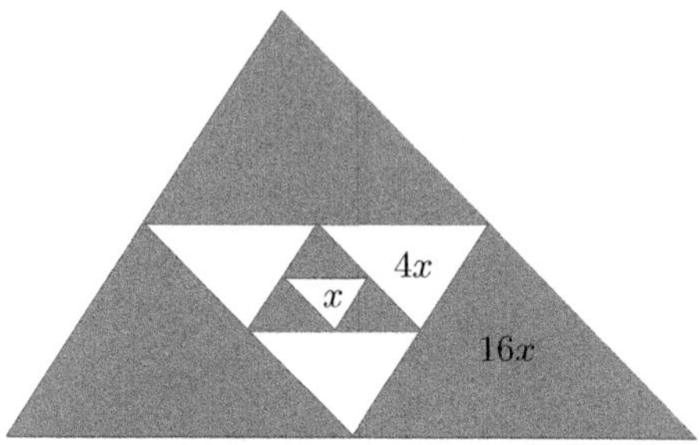

There are 4 small triangles (3 grey and 1 white), 3 medium triangles (all white) and 3 large triangles (all grey). 4 small triangles combine to make 1 medium triangle and 4 medium triangles combine to make 1 large triangle. So if we let the area of a small triangle be x, then the total grey area is $x + x + x + 16x + 16x + 16x = 51x$, and the total white area is $x + 4x + 4x + 4x = 13x$. Thus the required ratio is $\boxed{51 : 13}$

7. The first few terms are $96, 240, 600, 1500, 3750, 9375, 23437.5, ...$, after which there are no more whole numbers. So the answer is $\boxed{9375}$

8. $\{k\} - \{k-1\} = 358$, so $k(k+1) - (k-1)k = 358$, so $2k = 358$, hence $k = \boxed{179}$

9. Consider the diagram below.

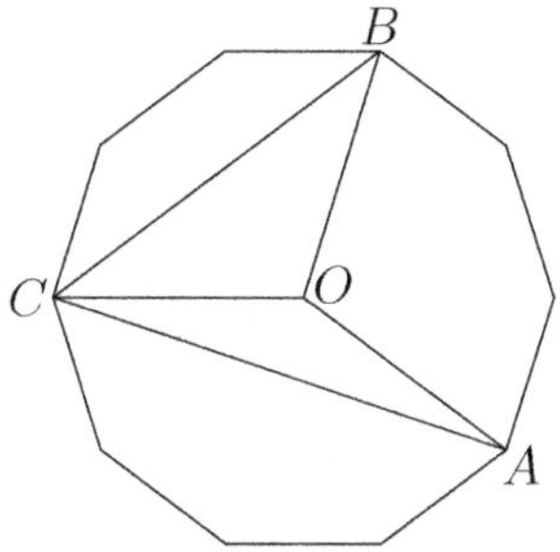

Let O be the centre of the decagon as shown. There are 4 edges around the perimeter of the decagon from A to C, so $\angle AOC$ is four tenths of $360°$, or $144°$. There are 3 edges around the perimeter from B to C, so $\angle BOC$ is three tenths of $360°$, or $108°$. $\triangle AOC$ is isosceles, so $\angle ACO = 18°$. Similarly, $\triangle BOC$ is isosceles, so $\angle BCO = 36°$. Hence $y = 18 + 36 = \boxed{54}$

10. The first few terms are $17, 3, 14, 11, 3, 8, 5, 3, 2, 1, 1, 0, 1, 1, 0, 1, 1, 0, \ldots$. We see that, after the first nine terms, the sequence just contains the block $1, 1, 0$ repeating over and over again. The sum of the first nine terms is 66, and from the tenth term to the 2013th term, the block $1, 1, 0$ appears 668 times completely; these 668 complete blocks sum to 1336. Finally, the 2014th and 2015th terms are both 1. It follows that the sum of the first 2015 terms is $66 + 1336 + 2 = \boxed{1404}$

11. A number is divisible by 36 if and only if it is divisible by 4 and 9. This requires the number formed by the last two digits to be a multiple of 4, and also the sum of the digits to be a multiple of 9. Since $1+2+3 = 6$, we need another 3 from somewhere. In fact, one digit 3 is the only extra digit required, since 1332 is a multiple of 36. So the answer is $\boxed{1332}$

12. Call two of the angles x and $2x$. Then the third angle must be $3x, \frac{x}{3}, 6x$ or $\frac{2x}{3}$. Considering each of these cases in turn, and using the fact that the interior angles in a triangle sum to 180°, we obtain four possible angle sets: $(30, 60, 90)$, $(18, 54, 108), (20, 40, 120)$ and $(32\frac{8}{11}, 49\frac{1}{11}, 98\frac{2}{11})$. Hence the complete solution set is
$\boxed{18, 20, 30, 32\frac{8}{11}}$

13. Let x denote the side length of one square. Since the area and perimeter are numerically equal, we have $5x^2 = 12x$, so $5x = 12$, and $x = \frac{12}{5}$. Hence the area is $5x^2 = \frac{144}{5}$ or $\boxed{28.8}$

14. Since 1 is less than nine times its cube root, the answer is positive. Let the whole number be x^3. We have $x^3 < 9x$, so $x^2 < 9$, so $x < 3$, so $x^3 < 27$, so $x^3 \leq 26$. Hence the answer is $\boxed{26}$

15. By dividing each bracket by $(1+2+3+\cdots+10)$, the fraction simplifies to $\frac{1\times 3}{2\times 4}$, which is equal to $\frac{3}{8}$, or $\boxed{0.375}$

16. Let x and y denote lengths as shown below.

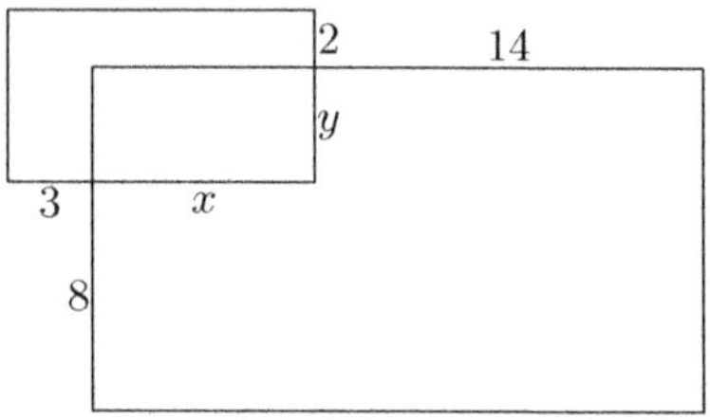

Since the dimensions of the larger rectangle are twice those of the smaller rectangle, we obtain two equations. Comparing widths gives us $x+14=2(x+3)$ and comparing heights gives us $y+8=2(y+2)$. Solving these yields $x=8$ and $y=4$. So the area of the small sheet is $11\times 6=66$, the area of the large sheet is $22\times 12=264$, and the area of the overlap is $8\times 4=32$. Hence the total area covered is $66+264-32=\boxed{298}$

17. Joe does not score full marks. Scoring 1/2 gives 50%, 2/3 gives 67%, 3/4 gives 75%, 4/5 gives 80%, and 5/6 gives 83%. However, 6/7 gives the required 86%, so the smallest possible number of marks obtained is $\boxed{6}$

18. If Charlie is telling the truth, then Bob is lying. If Charlie is lying, then Arthur and Bob are both telling the truth (which is impossible) or both lying. So we know for certain that Bob is lying. Hence Bob has at most 1 marshmallow, which he eats. So Bob has $\boxed{0}$ marshmallows left.

19. If $\frac{1}{n} > 0.006 = \frac{6}{1000}$, then $n < \frac{1000}{6} = 166\frac{2}{3}$. Similarly if $\frac{1}{n} < 0.007 = \frac{7}{1000}$, then $n > \frac{1000}{7} = 142\frac{6}{7}$. So the possible values of n are

$$143, 144, 145, ..., 165, 166,$$

that is $\boxed{24}$ numbers.

20. Every interior angle is less than 180°, so the most edges Johnny's polygon could possibly have is 12: 2 vertical edges on either side of the grid, 5 edges above them and 5 edges below them.

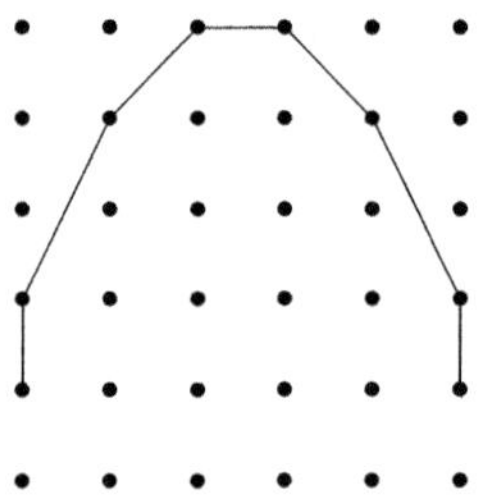

But if there are 5 edges above the vertical edges, as in the grid above, then there is not enough room below them to fit even 4 more edges, let alone another 5. If we remove one vertical edge, the best we might be able to do is 11, but again there is not enough room. So the best Johnny can hope for is $\boxed{10}$ edges, which is possible.

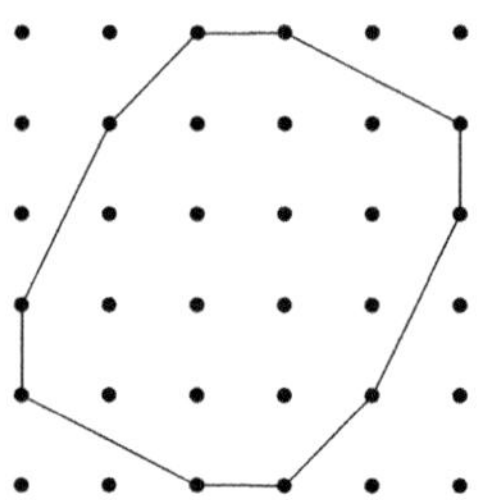

21. In 90 minutes, the cold tap can fill 3 bathtubs and both taps together can fill 5 bathtubs. So in 90 minutes the hot tap can fill 2 bathtubs, or 1 bathtub in $\boxed{45}$ minutes.

22. Consider the diagram below.

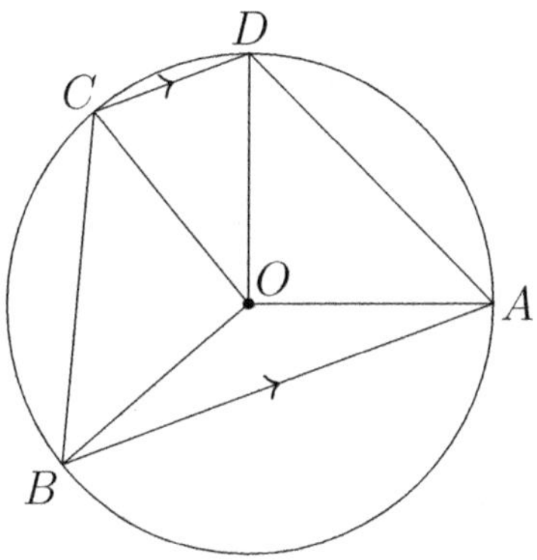

$\triangle AOD$ is right-angled and isosceles, so $\angle OAD = \angle ODA = 45°$. AB is parallel to CD, so $\angle BAD + \angle CDA = 180°$, so $\angle CDO = 68°$.

$\triangle ABO$ is isosceles, so $\angle OBA = 22°$ and $\angle AOB = 136°$. $\triangle CDO$ is isosceles, so $\angle OCD = 68°$ and $\angle COD = 44°$.

This means that $\angle BOC = 90°$, so that $\angle OBC = 45°$. Hence $\angle ABC = 45° + 22° = \boxed{67°}$

23. In 10 minutes, Raam walks $\frac{2}{3}$ km, so at 2.10pm he is $2\frac{1}{3}$ km from Tony's house. It remains to discover how long it takes Raam and Tony to cover $2\frac{1}{3}$ km between them. Since they can cover 10km between them in 60 minutes, they can cover $2\frac{1}{3}$ km between them in $\frac{60}{10} \times 2\frac{1}{3} = 14$ minutes.

    Hence they meet at $\boxed{2.24\text{pm}}$

24. Let x denote the smallest of the last nine numbers, so that the complete set of numbers is $(x-12), (x-11), (x-10), \ldots, (x+7), (x+8)$. The condition tells us that $(x-12)+(x-11)+\cdots+(x-1) = x+(x+1)+\cdots+(x+8)$. This simplifies to $12x - 78 = 9x + 36$, so $3x = 114$ and $x = 38$. Thus the largest number in the set is $\boxed{46}$

25. Suppose there are x rows and y columns in total. There are $175 + 400 = 575$ people not in Evariste's row, so $(x-1)y = 575$, that is $xy-y = 575$ (*). There are $312 + 264 = 576$ people not in Evariste's column, so $x(y-1) = 576$, that is $xy-x = 576$ (**). Comparing equations (*) and (**), we see that x must equal $y-1$, so $x^2 = 576$, hence $x = 24$ and $y = 25$. Given that 175 people are in front of him, and $175 = 25 \times 7$, he must be standing in $\boxed{\text{row 8}}$. Finally, given that 312 people are to his left, and $312 = 24 \times 13$, he must be standing in $\boxed{\text{column 14}}$

26. If $75n$ is a cube, say $75n = k^3$, then k^3 is divisible by 5. Since 5 is prime, k itself must be divisible by 5. Checking small cubes of multiples of 5, we find that $15^3 = 3375 = 75 \times 45$. So the smallest possible value of n is $\boxed{45}$

27. With two digits removed, we now have a number system with only eight digits. The number of three-digit numbers in this new system (allowing leading zeros) is $8 \times 8 \times 8 = 512$. That is, there are 512 numbers in the list $000, 001, 002$, ..., $998, 999$. So 1000 is the 512th number in Ashley's list, and 2000 is the 1024th number in his list. From this point, we can simply count back to the 1000th number, which is $\boxed{1960}$

Chapter 8

Diamond Solutions

1. Since $9 = 8+1 = 7+2 = 6+3 = 5+4$, we can't choose both 8 and 1, we can't choose both 7 and 2, we can't choose both 6 and 3 and we can't choose both 5 and 4. So we must get rid of at least four numbers, and the best we can hope for is $k = 6$. This is achievable: both $(2, 3, 4, 8, 9, 10)$ and $(2, 3, 5, 8, 9, 10)$ work. Hence the maximum possible value of k is $\boxed{6}$

2. In the first sequence, there are four dots in the first figure, and to obtain the next figure, one dot is added to each side. Hence the nth figure contains $4n$ dots. In the second sequence, there are six dots in the first figure, and to obtain the next figure, one dot is added to each edge (of which there are seven). Hence the mth figure contains $7m - 1$ dots. This means we are looking for a number greater than 2017 that is both divisible by 4, and 1 less than a multiple of 7. Checking multiples of 4 one by one, we find that none of 2020, 2024, 2028 and 2032 work, but 2036 is good. So the answer is $\boxed{2036}$

3. Any number of the form $1000\cdots005$ is divisible by 3 (since the digits add to 6) and by 5 (since the last digit is 5), and so is divisible by 15. Hence $\boxed{15}$ is elastic.

   Any number of the form $1000\cdots008$ is divisible by 2 (the last digit is even) and by 9 (the digits add to 9), and so is divisible by 18. Hence $\boxed{18}$ is elastic.

   Finally, any number of the form $4000\cdots005$ is divisible by 5 (the last digit is 5) and by 9 (the digits add to 9), and so is divisible by 45. Hence $\boxed{45}$ is elastic.

4. The scores 1-1, 2-2, 3-3, 4-4, 5-5, 6-6, 7-7 and 8-8 must all occur. So we are free to pick the scorer of the odd-numbered goals (the first goal, the third go, and so on...), but then the scorers of the even-numbered goals are fixed. The final scorer must be S since they won 9-8. Hence, the total possible number of goal sequences is $2^8 = \boxed{256}$

5. Anna can guarantee a win.

   Here is one possible winning strategy. Anna takes 5, leaving 25. Daniel must then leave 21 or 22. Anna takes 5, leaving 16 or 17. Daniel must then leave 12, 13 or 14. Anna takes 5 leaving 7, 8 or 9. Daniel must then leave 3, 4, 5 or 6. Finally, Anna takes the most she can, leaving Daniel with 0, 1 or 2 beans, so that he cannot move.

   Alternatively, consider that at some point, Anna will be left with a pile containing 2, 3, 4, 5, 6, 7, 8, 9 or 10 beans (since the most beans the two of them can remove with one turn each is 9). Checking each of these cases one by one we find that Anna can guarantee victory from any of these positions.

6. Since $3n$ has an odd number of factors, it is a square number. Searching through small squares divisible by 3, we find that 81 has exactly 5 factors. This would make $n = 27$, and $4n = 108$, which has 12 factors. In fact, $3n$ must equal $3^4 = 81$, since any number with exactly 5 factors must be of the form p^4, where p is some prime number.

7. Here are two arrangements of trees that permit $T = 5$; that is, in both arrangements, there are five possible tree counts depending on where Helen stands.

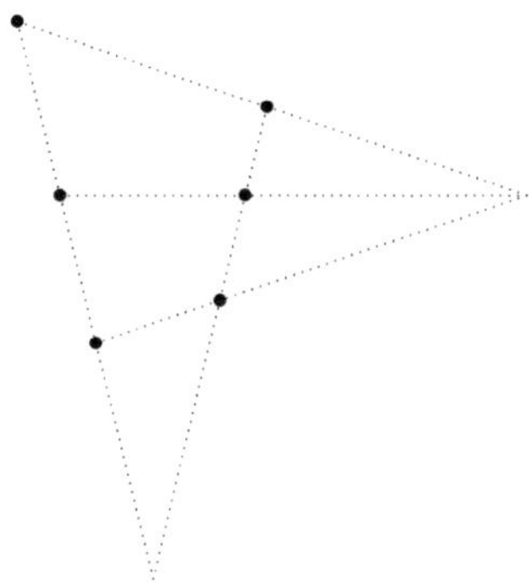

Suppose that $T = 6$ were possible. Then on some day, only one tree can be seen. This means that all the trees are in a straight line. But in this case, it is impossible to stand anywhere and see just 3 trees. It follows that the greatest possible value of T is $\boxed{5}$

8. If all the remainders are the same, then they must all equal $0, 1$ or 2. If they are all 0, then the number is a multiple of $3 \times 4 \times 5 = 60$. There are 16 such numbers: $60, 120, 180, ..., 960$. If all the remainders are 1, then the number is 1 more than a multiple of 60. There are 17 such numbers: $1, 61, 121, ..., 961$. Finally, if all the remainders are 2, then the number is 2 more than a multiple of 60. Again there are 17 such numbers: $2, 62, 122, ..., 962$. So in total there are $16 + 17 + 17 = \boxed{50}$ numbers with required property.

9. Consider the diagram below.

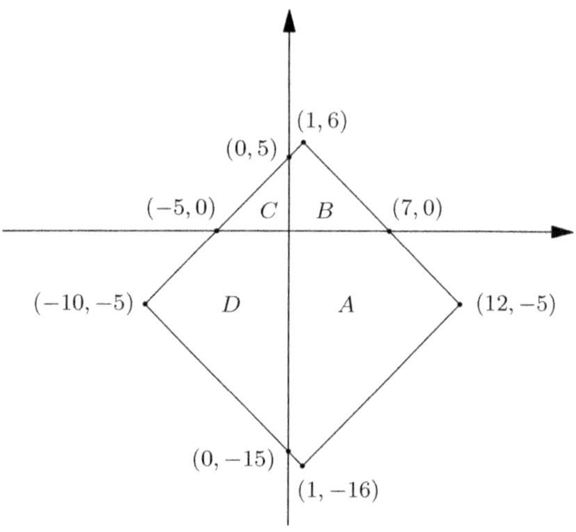

The coordinates of the intersections with the axes can be calculated by noticing that, on the perimeter of the square, for every unit moved right or left, you move one unit up or down. Now considering the areas of triangles, we obtain $C = \frac{1}{2} \times 5 \times 5 = 12.5$, $B + C = \frac{1}{2} \times 12 \times 6 = 36$, and $C + D = \frac{1}{2} \times 20 \times 10 = 100$. The area of the entire square is half the product of its diagonals, that is $\frac{1}{2} \times 22 \times 22 = 242$. Hence, the required area is $A = 242 - 36 - 100 + 12.5 = \boxed{118.5}$

10. The largest possible value of T is 66, which occurs when the split is $\{2, 3\}$ and $\{4, 5, 6\}$. The smallest possible value of T is 17, which occurs when the split is $\{2, 3, 4, 6\}$ and $\{5\}$. Hence the answer is $66 - 17 = \boxed{49}$

11. The cool numbers between -300 and 300 are $-299, -296, -293, \ldots, 295, 298$. These can be paired up: $(-299, 298), (-296, 295), \ldots, (-2, 1)$. In each pair, the sum is -1, and there are 100 pairs, so the required sum is $\boxed{-100}$

12. Every odd number n can be written in the form $n = 2k + 1$ for some whole number k.

    Notice that n lies exactly halfway between 6 and $2n - 6$, where 6 is a multiple of 3 and $2n - 6 = 2(2k + 1) - 6 = 4k - 4 = 4(k - 1)$ is a multiple of 4. This proves the result.

    Alternatively, any number n lies exactly halfway between $-6n$ and $8n$, where $-6n$ is a multiple of 3 and $8n$ is a multiple of 4. $\square$

13. Consider the diagram below.

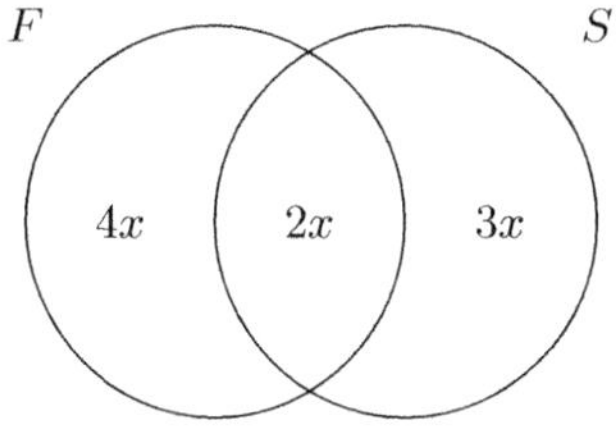

Let $4x, 2x$ and $3x$ denote respectively the number of students taking French only, both languages and Spanish only. This obeys the conditions given in the problem. When six double-linguists drop French, the numbers change to $4x, 2x-6$ and $3x+6$, in particular $6x-6$ students in total study French.

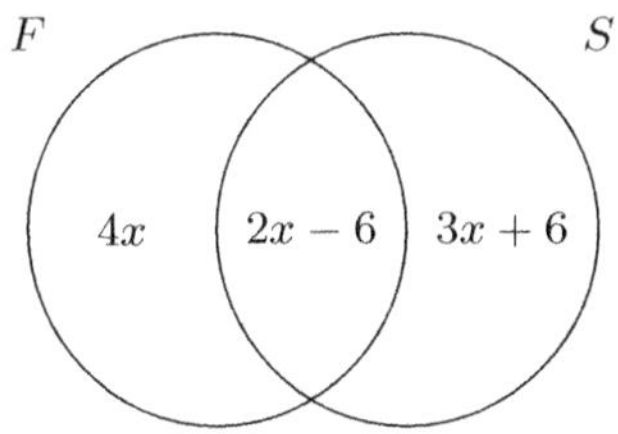

After the change, only a quarter of the French students study Spanish, so $6x-6 = 4(2x-6) = 8x-24$, so $2x = 18$ and $x = 9$. Thus, the total number of students at the academy is

$4x + 3x + 2x = 9x = \boxed{81}$

14. Consider the diagram below.

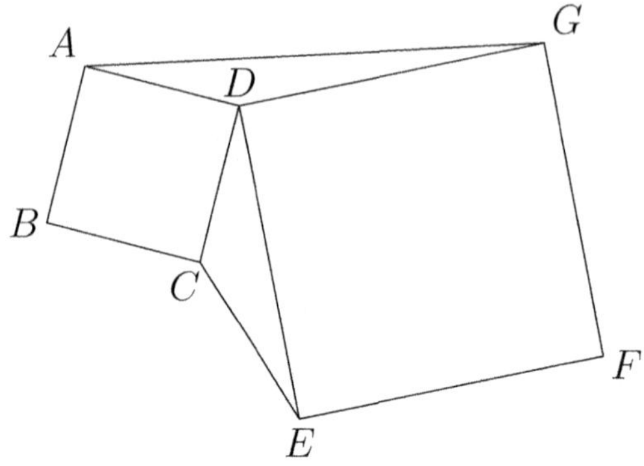

$\angle ADC + \angle EDG = 180°$, so $\angle ADG + \angle CDE = 180°$. Also, $DE = DG$. This means that it is possible to rotate $\triangle CDE$ anticlockwise about D so that DE aligns with DG.

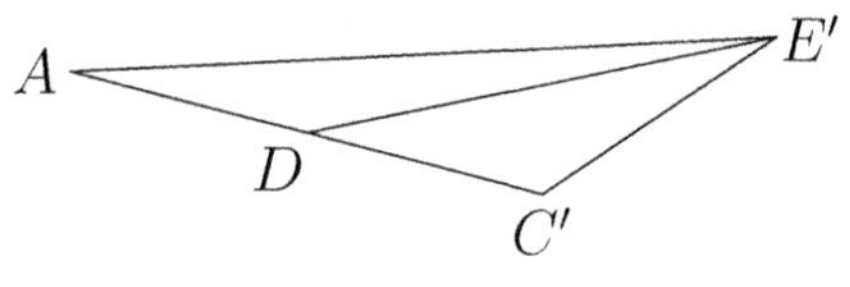

Now we have two triangles with the same base and the same height, hence they have the same area, as required. □

15. Let x be the number of Factory A workers producing robots, and let y be the number of Factory B workers producing robots. This means that $20 - x$ Factory A workers and $24 - y$ Factory B workers will be producing stuffed animals. So in 60 minutes, A will produce $30x$ robots and $20(20 - x)$ stuffed animals, and Factory B will produce $20y$ robots and $15(24 - y)$ stuffed animals. So we require

$$30x + 20y = 20(20 - x) + 15(24 - y)$$

This simplifies to $10x + 7y = 152$. Given that x and y are whole numbers, this equation has very few solutions. Indeed, we require $7y$ to end with the digit 2. Thus our only options are $(x, y) = (11, 6)$ or $(4, 16)$. In the first case, the number of robots produced is $30 \times 11 + 20 \times 6 = 450$, and in the second case, the number of robots produced is $30 \times 4 + 20 \times 16 = 440$. So the optimal solution sees 11 Factory A workers and 6 Factory B workers producing robots, and all other workers producing stuffed animals. In this scenario, $450 + 450 = \boxed{900}$ toys are produced in total.

16. There are $9 \times 10 \times 10 \times 10 \times 10 = 90000$ five-digit numbers. Of these, $4 \times 5 \times 5 \times 5 \times 5 = 2500$ contain only even digits, and $5 \times 5 \times 5 \times 5 \times 5 = 3125$ contain only odd digits. It follows that the number of five-digit numbers that contain an even and an odd digit is $90000 - 2500 - 3125 = \boxed{84375}$

17. Number the seats left to right from 1 to 8. The four members of the Newton family can sit in positions $(1, 2, 3, 4), (5, 6, 7, 8), (1, 2, 5, 6),$ $(3, 4, 7, 8), (1, 2, 7, 8)$ or $(3, 4, 5, 6)$.

    Once the Newtons choose their set of four seats, the Leibnizes occupy the other four seats. For each of the six choices above, there are $4 \times 3 \times 2 \times 1 = 24$ ways for each family to arrange themselves within their assigned seats. Hence the total number of possible arrangements is

    $6 \times 24 \times 24 = \boxed{3456}$

18. The number of ways of choosing three different points from a set of nine points is $\frac{9\times8\times7}{3\times2\times1} = 84$. But some of these choices involve three points on the same straight line, which is no good. There are 10 ways to choose three points from the horizontal line, and 10 ways to choose three points from the vertical line. So the final count is

$84 - 10 - 10 = \boxed{64}$

19. We have $10^3 - 1 = 999$ and $10^6 - 1 = 999\,999$. Since $999\,999 = 1001 \times 999$, the numbers we are counting are all of the form $999d$, where d is a factor of 1001. Since $1001 = 7 \times 11 \times 13$ has eight factors $(1, 7, 11, 13, 7 \times 11, 7 \times 13, 11 \times 13$ and $7 \times 11 \times 13)$, the answer is $\boxed{8}$

20. A number is a multiple of 4 precisely when the number formed by the last two digits is a multiple of 4. This property must hold for Ashley's original choice and its reverse. So the maximum possible value of his number is 8896 and the minimum is 2104, and the answer is

$8896 - 2104 = \boxed{6792}$

21. Add in the lines OB, BC and CD as shown.

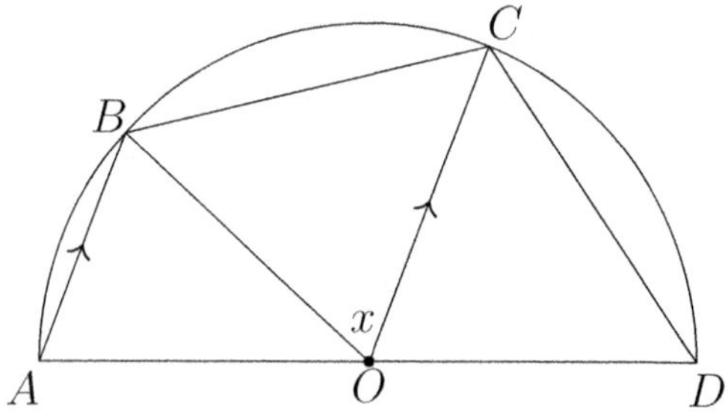

Let $\angle BOC = x$.

By alternate angles, $\angle ABO = x$. Since $\triangle AOB$ is isosceles, $\angle BAO = x$. By corresponding angles, $\angle COD = x$. Now consider triangles $\triangle BOC$ and $\triangle COD$. In both triangles, two radii enclose an angle of size x. Hence $\triangle BOC$ and $\triangle COD$ are congruent (triangles with the same size and shape), and thus the distances BC and CD must be equal. $\square$

22. A number is solar if it is divisible by 5 but not by 2 or 3. This property is held precisely by numbers of the form $30n + 5$ and $30n + 25$, that is, whole numbers that leave a remainder of 5 or 25 upon division by 30. The set of such numbers less than 2016 is $5, 25, 35, 55, ..., 1985, 2005, 2015$, which is $\boxed{135}$ numbers in total.

23. Let us look for a starting number x so that after just one press of each button, the final number is again x. We require $-2(x - 1) = x$, so $-2x + 2 = x$, so $3x = 2$, and $x = \frac{2}{3}$. But if $\frac{2}{3}$ remains after one press of each button, it must remain after exactly ten presses of each button. So $n = \boxed{\frac{2}{3}}$

24. Let A be Anthony's speed, and B be Benjamin's speed. Since 3 minutes is 180 seconds, the information given implies that

$$180(A - B) = 40(A + B).$$

This simplifies to $7A = 11B$, and so $\frac{A}{B} = \boxed{\frac{11}{7}}$

25. Suppose there was a Team A and a Team B. There are $\frac{6\times5\times4}{3\times2\times1} = 20$ ways to choose the three boys to go in Team A, and there are 20 ways to choose the three girls to go in Team A. That is 400 choices for the members of Team A; then Team B is fixed. But in the original problem there was no Team A and Team B, so we have counted every possibility twice. Thus the correct total is $\boxed{200}$

26. Add the line AD to the diagram as shown below.

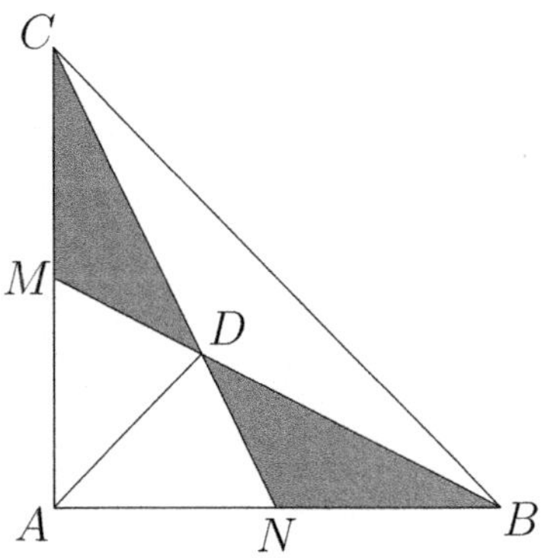

Let [...] denote area.

We have $[ANC] = \frac{1}{2} \times 5 \times 10 = 25 = [ABM]$ and so, since the region $ANDM$ is common to $\triangle ANC$ and $\triangle ABM$, it follows that the two grey triangles also have equal areas, say $[CDM] = [BDN] = x$.

Now, $\triangle AND$ and $\triangle BND$ have the same base and same height, so they have the same area x. Similarly, $\triangle ADM$ and $\triangle CDM$ have the same area x. So we have $3x = [ANC] = 25$, hence $x = \frac{25}{3}$. Thus the area of the grey region is

$2x = \boxed{\frac{50}{3}}$

27. Consider the diagram below.

1	2	3
4	5	6
7	8	9

Without any restrictions, there are $2^9 = 512$ different 3-by-3 square patterns. Of these 512, there are $2^3 = 8$ patterns with order 4 rotational symmetry (we are free to choose the colour of tiles 1, 2 and 5). For patterns with order 2 rotational symmetry, note that we are free to pick the colour of tiles 1, 2, 3, 4 and 5, which yields $2^5 = 32$ patterns; however, 8 of these have order 4 rotational symmetry, and have already been counted above. So there are $32 - 8 = 24$ patterns with order 2 rotational symmetry. Finally, the remaining $512 - 8 - 24 = 480$ patterns have no rotational symmetry.

But two patterns are considered the same whenever one is a rotation of the other. So the 24 patterns with order 2 rotational symmetry contain just 12 different patterns, each appearing twice. Similarly, the 480 patterns with no rotational symmetry in fact contain only 120 different patterns, each appearing 4 times. Thus, the number of different patterns (as far as this problem is concerned) is $8 + 12 + 120 = \boxed{140}$

Chapter 9

Final Thoughts

Favourites

Here are my favourite problems in the collection.

GOLD 10 *The sequence of differences*

A little investigation that rewards determination.

GOLD 12 *Angles in a triangle*

A simple premise, but difficult to find every solution.

GOLD 17 *Joe and the test*

An unusual problem about test scores.

DIAMOND 3 *Elastic numbers*

The puzzle this book is named after.

DIAMOND 7 *Helen on the island*

My single favourite problem in the collection, inspired by a glass in a restaurant.

DIAMOND 23 *Clare and the button*

Multiple solutions, one horribly complicated, one devilishly simple.

To find out more...

I hope you enjoy the problems and find the written solutions helpful. If there is something you want further explained, send a query to

elasticnumbers@gmail.com

You can also send in alternative solutions to problems, or tell me which puzzles were your personal favourites. For discussions on some fantastic problems from around the world, visit my blog at

puzzlecritic.wordpress.com,

and you can follow me at @puzzlecritic.

Never stop learning

Printed in Great Britain
by Amazon

62146035R00068